LET'S FIND OUT ABOUT THE SUN

BY MARTHA AND CHARLES SHAPP
Pictures by Stephanie Later

Revised full color edition

FRANKLIN WATTS, INC.
NEW YORK, N.Y./1975

What a beautiful day!
The sun is shining and everything looks
 bright.
It's nice and warm out.

Can you feel the heat of the sun? The sun that gives this heat is very far away.

It is hard to believe, but the sun is millions and
    millions of miles away.
It's hard even to imagine how far away that is.
Suppose a jet plane were to fly to the sun.
Suppose the plane could fly 600 miles an hour.

If the plane flew day and night without stopping, it would take nearly 18 years to get to the sun.

The sun doesn't seem very big to us.
But it is really enormous.
One million balls as big as our earth could fit inside the sun.
The sun looks small because it is so far away.
Do you feel hot on a summer day when the temperature is one hundred degrees Fahrenheit (38 degrees Centigrade)?
The temperature on the sun is over 10,000 degrees Fahrenheit (5,500 degrees Centigrade)!

Our earth is cool enough to live on because the hot sun is so far away.
If the sun were much nearer, our earth would be too hot to live on.
If there were no sun, the earth would be so cold that nothing could live on it.
Aren't we lucky to be just about the right distance away from the sun?

The earth circles the sun.
It takes one year for the earth to go completely around the sun.

And at the same time it is going around the sun, the earth itself is spinning around like a top. The earth spins around once every 24 hours.

But while the earth is circling and spinning, its movement is smooth and steady.
It is so smooth and steady that you cannot feel you are moving.
That's why people long ago thought that the earth was standing still and the sun was moving.
They thought that the sun traveled across the sky each day.

In the morning the sun seems to rise in the east.
As the day goes on, the sun seems to travel across the sky.
In the evening the sun seems to sink in the west and then it is night.

The sun lights half the earth at a time.
It is day on the lighted side.
It is night on the dark side of the earth.
Almost every part of the earth turns from
   day to night during every 24 hours.

If you have a globe you can do an experiment
that shows how day turns into night
on the earth.
Use a flashlight for the sun.
In a dark room, shine the make-believe sun
on the globe.
Spin the globe slowly.
What happens to different parts of the globe?

Did you ever notice your shadow at different times of the day?
Shadows change as the day goes on.
In the morning the sun makes long shadows.
In the middle of the day when the sun seems to be right over your head, your shadow is very, very small.
In the afternoon as the sun goes down there are very long shadows again.
But the long afternoon shadows fall a different way from the long morning shadows.

Long, long ago people told time by the shadows. They made sundials.
You can make a sundial and tell time by shadows.
Put a piece of paper in the sunlight.
Stand a pencil in a spool in the middle of the paper.
The shadow of the pencil will fall on the paper.
The pencil's shadow will move as the day goes on.
Use a watch and every hour mark the place where the shadow falls.

Now your paper is a sundial and you can tell time every sunny day.
Just look at your paper sundial and see where the shadow falls.

You can find out for yourself what happens to a plant that gets no sun.

Put a healthy green plant in a dark closet. Water it every day but don't take it out of the closet.

You will see it get paler and paler and lose more and more of its leaves.

And if it stays in the dark too long, it will die.

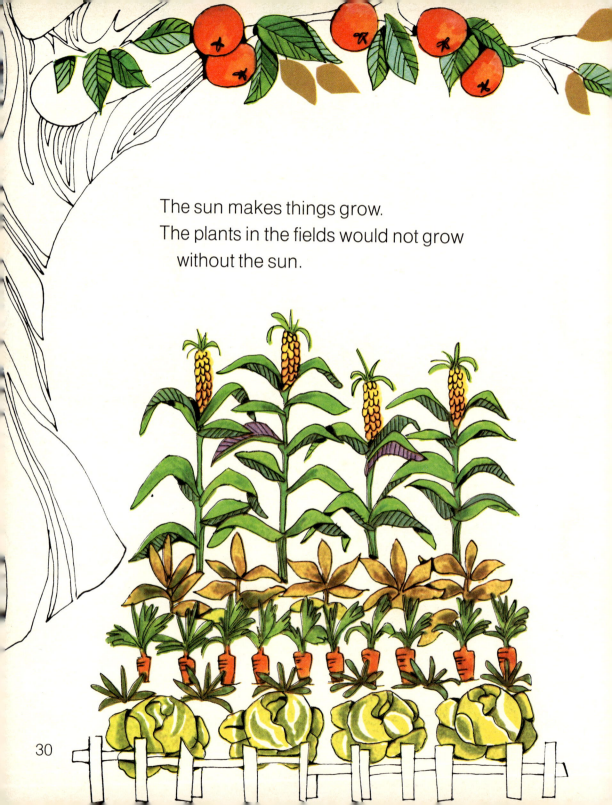

The sun makes things grow.
The plants in the fields would not grow without the sun.

Animals need the sun.
Animals live by eating the plants that grow in the sun.

People need the sun.
We eat the plants that grow in the sun.
We eat the animals that ate the plants that
   grew in the sun.

The sun gives us many things we need.
Many of our clothes are made of cotton.
The cotton was once a plant that grew in
   the sun.

Our sweaters are made of wool.
The wool comes from sheep that grow by eating plants that grow in the sun.

The wood we use to build a house was once a tree that grew in the sun.

The sun makes our earth beautiful.
It gives us light.
It keeps us warm.
It makes things grow.
We could not live without the sun.